WIRELESS WORLD

The Earth with Electronic Waves Around Us

EMORY CLARK

ONE

How It All Started

It was February 19th, 1880, that the first ever wireless telephone communication device was invented by Alexander Graham Bell and his assistant Charles Summer Tainter in Washington, D.C. On June 3rd of the same year, Bell's assistant conveyed a wireless message from the top of Franklin school to the window of Bell's lab. This is regarded as the first ever wireless conversation in the history of the world. This device was termed a "photophone" because the audio was sent through a beam of light. Bell regarded it as one of his greatest inventions, even greater than the telephone.[1] The photophone required sunlight to send messages and also a direct sight between the receiver and the transmitter. For these reasons, its viability for daily use was low. But it showed a path to a revolutionary change in the field of communication.

And here we are! We're living in an age where there are all sorts of waves surrounding us every day; every minute even. Look around you. Can you imagine a life without wireless devices? Life as you know it would not be possible.

You don't need to go far back in time to see what you'd be losing should wireless technologies not be present in your life. As early as thirty years ago there were no smartphones, no tablets, no portable computers that you could use anywhere you wanted. The internet was still only available to a handful of people and checking news or watching movies online was a far-flung dream.

Almost everything we use in our day-to-day lives has been revolutionized due to wireless technology. Your TV, air conditioner, your phone, headphones, internet, keyboard, and mouse; everything is blessed with it. Text messaging your best friend, watching exciting series on Netflix, and submitting your assignment at the last minute requires access to wireless technology in this era. Wireless technology has helped us to forget all limitations, regardless of location, and has paved the path of ultimate convenience. This technology developed so fast in the last 50 years that if you look back you would be amazed at the lives people lived back then. In today's world, it is hard to think about living without wireless technology but yet our not-so-distant ancestors lived quite happy, fulfilling lives without it. Even though people have been using it for the entirety of their lives, only few have an idea of how it really works.

What is Wireless Technology and How Does It Work?

Wireless communication means information transferring between two or more devices or points that do not use an electrical conductor as a medium. (Wikipedia) [2] Before learning about wireless technology, we need to learn about the areas where it is applied. Information from a wireless device can be unidirectional, the same as radio and TV broadcasting. It can also be bidirectional, like information transmitted via phone calls or video calls.

To transfer data without wires, this technology uses signals or electromagnetic waves. Now how are the waves created in the first place? You need two network infrastructures that transfer information between them, the transmitter and the receiver. It all starts with a transmitter, in which an oscillator generates waves. These waves are propagated by internal wires to an antenna. The antenna converts the electric current to an electromagnetic wave and propagates it. A signal does not usually go straight to a receiver from the antenna. The antenna pushes the electromagnetic wave in different directions. The waves may find thousands of obstructions along the way but they are not destroyed. They still manage to reach the receiver. At the receiving end, the opposite thing happens. The receiver receives the electromagnetic waves and converts them back to a sound or video which can be heard or watched by a receiver. [3]

Let me give you an easy example. Suppose you are at a radio station and your voice is collected by a microphone that has an electrical connection. These sound waves create vibrations and are transmitted to the antenna. This antenna propagates all the information in the form of electromagnetic waves or radio waves. Now, a radio acts as a receiver. So, if you have a radio, you can receive the same electromagnetic waves. The radio itself will convert the wave into a sound and thus you are able to hear your own voice.

The History of Wireless Technology

Wireless technology is not something new. Scientists have been trying to invent newer ways to communicate without wires. After the invention of the "photophone" by Bell and his assistant, many wireless signaling technologies were available. These included the patented induction system by Thomas Edison that allowed telegraphs on a running train to connect with telegraph wires running parallel to the tracks. They also included William Preece's induction telegraph system of sending messages across water bodies. Although they were all great ideas, none of them became a commercial success.

James Clerk Maxwell showed theoretically that it is possible that electromagnetic waves could radiate through free spaces. This theory excited the minds of many scientists. In 1888, Heinrich Rudolf Hertz was able to prove Maxwell's theory with an experiment that transmitted airborne electromagnetic waves. Later, these waves were termed as "Hertzian waves" and many scientists and inventors started experimenting with them. After several years of experimenting and trial, Italian inventor Guglielmo Marconi invented the first complete telegraphy system, based on Hertzian waves (radio transmission). [4]

The Invention of Radio

Think about radio. What comes to mind? Probably crystal-clear music or the voice of your favorite musician or RJ, broadcasted from a radio station. But radio wasn't always like this. It took years of development and hundreds of scientists to build what we have today.

Although numerous scientists were working on the Hertzian waves, they could not build a complete, commercially successful model. Marconi, on the other hand, read through the works of others who were experimenting with the waves and was successful in developing devices such as portable transmitters and receivers that could work long distance. In August 1895, Marconi could only transmit signals up to half a mile between locations. Later, he raised the height of the antenna he used and grounded his transmitter and receiver. This action led to improvements in the system and now the radio was able to transmit signals up to 3.2 km, even over hills. In 1901, Marconi became the first person to send wireless signals successfully across the Atlantic, from Cornwall England to St. Johns, Newfoundland. Marconi received the Nobel Prize in Physics for his work in 1909. Radio communication was already commercially used in 1900.

Initially, radio technology was only used to contact ships using the morse code. In December 1906, the first music signals were transmitted from Brant Rock by Canadian experimenter, Reginald Fessenden. This began a great opportunity and a new era for commercial voice and music broadcasts. Years went by and the radio technology started to move further forward. In the 1920s, people started to make all sorts of things with radio technology and setting up different broadcasting stations. The first ever broadcast took place in November 1920. Pittsburgh's KDKA station transmitted this broadcast on election day. Thanks to the radio, people were able to hear the result of the presidential election without reading about it in the newspapers the next day.

One of the initial problems that radio faced was poor audio quality and interference that resulted in distorted voice and audio signals. Those issues were later solved by the invention of frequency modulation (FM), which was able to provide better audio quality compared to amplified modulation. [5]

From Audio to Video

Ever thought how you're able to watch videos on many platforms so smoothly? There is so much more to know about how it came to being in the first place. Television has greatly evolved throughout the last few decades. What was once a commodity for the elite has now become easily accessible to everyone.

In the 1920s, radio was the most important medium of entertainment for most people. But soon people started craving for images to go along with the voice and music. A significant milestone in the industry of television was the invention of electronic television, which was demonstrated by Philo Taylor Farnsworth in 1927.[6] Taylor was only 21 years old at that time and he grew up until the age of 14 in a house which had no electricity. While in high school, he started to develop a system that was able to capture moving images in a way that could be coded into radio waves and can be converted back to moving images. This was similar to the mechanism of radio but it allowed video conversion. Many scientists worked on similar concepts but Farnsworth's invention of scanning images with a beam of electrons is the true ancestor of modern-day television. The first image Farnsworth transmitted was a simple line. As time passed, the concept of mechanical television was starting to become obsolete. By 1934, all televisions were converted to electric versions.

In the 1930s, however, the development of technology came to a halt due to the Great Depression. In the later part of the decade, however, the RCA invested $50 million in the development of more advanced versions of electronic television. At the 1939 World's Fair held in New York City, RCA decided to televise the ceremonies and broadcast the speech of President Franklin D. Roosevelt. He became the first president to appear on television.

The Second World War halted the development and improvement of television and radio remained the most popular wireless technology. In 1946, TV sets went for sale. Initially less popular because of the price tag, slowly their popularity started to rise and middle-class families were able to afford TVs in their households. Over 2 million families in the U.S had televisions at the end of 1950.[7] The spread of TV sets to average families led to better opportunities for TV stations, entertainment, and marketing. People were finally starting to realize and see that there were no boundaries to technology and that anything could be wireless. Television was undoubtedly a great step in wireless technology. Its invention led to greater developments in wireless technology.

Connecting the World

When you are clicking on a link in your browser, you are actually able to access information from another computer located far away, probably in another country or a different time zone. Yes, I am referring to the internet. The advent of the internet was a miraculous discovery. It has opened horizons and helped to convert this world into a big global village. You can now order anything, watch any videos, and learn from any articles, as long as they're made available online. Billions of computers are connected to each other in such subtle ways that you can use them without any sort of delay. We mostly take all this wireless technology for granted but if you look at it closely, it seems like a miracle. Moving forward from television, the internet is yet another invention that has started a new era to connect the whole world.

Here is how it all started. The invention of the computer was made quite a long time ago, during the Second World War. At that time, computers didn't do much other than solving complex math problems. They were huge, clunky machines that were developed for the purpose of solving these complex equations and problems. But even with the best computers, it would take months to solve a simple physics equation. An early step towards the development of the internet came when colleges started separating their computer terminals. A person was then able to write code on the computer. This made many people willing to experiment with the machine without handling the circuits and complex machinery. In a sense, it was similar to cloud computing as we use it today, except for the fact that cloud computing could not work without the internet. This led people to think about how to connect more than one machine together.

The Advanced Research Project Agency (ARPA) was founded in the U.S. to allow the country to stay one step ahead in its race with the Soviets. Joseph Licklider, a computer enthusiast, convinced ARPA to fund a project that was meant to create a computer network to connect scientists and engineers throughout the U.S. ARPA agreed and started building the network in 1969. They called it ARPANET. It started like a type of messaging service between different universities. It was the first of its kind. As it started to grow more and more, unique features started to be added to this computer network. One of the really great innovations of ARPANET was called "packet switching."

Before packet switching, when people needed to call someone else on the phone, they went through an operator. The operator would plug the circuit of one phone to that of another to make the right connection. That process is called "circuit switching." The internet cannot work in that way because if it did, you would only be able to connect to one computer at one time and it would take a lot of time to connect to another computer. Modern day internet helps you to connect to thousands of computers and all of them are immediately responsive to your requests. Circuit switching is useless in this scenario, which is why packet switching was invented, in which every computer could send messages with the same wires instead of getting a new one every time. It is similar to a parcel delivery where every message is a packet and it has an address label, which is a series of numbers representing a unique address.

When it was invented, the system wasn't as straightforward as it is now. The sender computer sent the packet (the message) to the nearest computer by looking up at the destination address. The second computer would do the same, and this process would repeat until the packet would finally reach the destination computer. ARPANET used phone lines for sending out these packages. It worked great initially but as more people joined the network, it became more difficult to operate. The process required every new computer to be updated on the initial computer's address list, which was difficult to accomplish fast.

As things started to get more complicated, ARPANET scrapped the packet switching system and selected Stanford University as the official record keeper of computer addresses in 1973. With the new system in place, ARPANET was able to grow its network to more and more locations including England and Norway.

By the late 1970s, there were a lot of new networks popping up everywhere. As they developed, they used different ways to send messages. This incompatibility needed a unifying solution and it soon got one. The solution was a set of programs. They were the transmission control protocol (TCP)/ internet protocol (IP), which we still use to this day. The transmission protocol was a standard way of formatting packets so that every network could be on the same page, while the internet protocol was the standard way of assigning addresses. With the advent of the TCP/IP protocol, connecting different computers became a lot easier. All the computers in the world that were wirelessly connected to each other could send information from one to another in a much more seamless fashion. This is what became known as the internet.

As time passed, the networks started to grow bigger and Stanford University was getting overloaded with the amount of address records it needed to keep. Sometimes, its list had errors which would then create problems throughout the whole network. Sending emails was becoming a real headache. Originally spelled "e-mail," the word was invented in 1971. Within two years, three out of four packets sent through the ARPANET network were emails. Different computers had different email programs. Some computers needed to provide a complete list of the addresses between the sender and the receiver, so people had to keep an updated version of the Stanford list. They had to type out the entire path before they could send an email. With hundreds and thousands of computers in the network, this became a really difficult challenge to sort out. The entire structure of the internet needed to change.

That is when ARPANET came up with the domain name system (DNS). Instead of separating each host and storing their address in a random order, the hosts were arranged in domains. The domain structure organized all the different types of hosts from over the world in a way that computers were able to handle. DNS added a whole new network, whose only job was to keep the addresses of the domains organized. So, if you wanted to send an email, you didn't have to map the whole path anymore. That became and is still to this date the DNS server's job. This means that if you are clicking a link in your browser, you are telling the DNS server to show you the information from the domain you selected. The DNS will do its job.

Short-Range but Powerful

There is a great chance that you're using Bluetooth in your day-to-day life. It is used for so many functions and components, such as your earbuds, keyboard, mouse, speakers, controllers, and so much more. It is a technology that gets better every year. When the technology was introduced, people were amazed at how fast they could send large amounts of data using it.

The Bluetooth standard was originally adopted by Dr. Jaap Haartsen from Ericsson in 1994. It was originally developed to replace wired communication by using short-range ultra-high frequency radio waves between 2.4 and 2.485 GHz. The first consumer Bluetooth technology launched in 1999. It was a handsfree headset. The Bluetooth 1.0 officially launched that year for chipsets, dongles, mouses, phones, and more.

The first version of Bluetooth was not without its problems. It had a lot of connection problems, as well as a comparatively low range of 10 meters. It was not really efficient. The 1.0 specification offered only 721 kbps as peak speeds. The best use of Bluetooth was probably in auditory products, but the standard version was not good enough for music at full bandwidth. The offered speed, 721kbps, couldn't cope with any highresolution audio even early on. This may have been because the technology first developed to allow wireless voice calls rather than listen to music.

Over the years the Bluetooth technology has developed to a much more effective version. But with the advent of super-fast internet and wireless services, people don't always require Bluetooth technology for transferring data. Yet it is a milestone in the field of wireless technology that you simply cannot overlook. [8]

Complementing the Internet

From its invention, Wi-Fi played an amazing role in keeping the world connected. The way Wi-Fi works is quite similar to a radio. Recent developments in Wi-Fi technology have helped to maintain high productivity, good health, and even a great social life.

Prior to 1997, there were several wireless technologies used for connecting devices. But vendors felt the need for a shared wireless standard, much like ethernet. So, vendors contacted the Institute of Electrical and Electronics Engineers (IEEE) to discuss the possibility of forming a brand-new standard committee for wireless communication. The new committee started their job in 1990 under the name of "802.11." The committee finally agreed on and released the first of the standards in 1997. Immediately after the signing of IEE802.11, major networking companies and computer manufacturers developed and brought to market Wi-Fi products. The first generation of Wi-Fi products had a maximum transfer rate of 2 Mbit/s and operated in 2.4 GHz band.

Technical standards are long documents and as people start implementing them, they are very often misinterpreted. That's what happened to the 802.11 standard. In 1999, six major vendors founded a new alliance to improve cross-compatibility. The new alliance pursued a good cause and it also wanted a good name. That is how it came up with "Wi-Fi." It was not meant to be an acronym. In fact, the word has no prior meaning at all. It's just a catchy word that sticks in your head when you hear it. The Wi-Fi Alliance has hundreds of vendors working together. But even now, when it comes to developing standards, the 802.11 committee is still in charge. [9]

There are many wireless technologies out there; they include wireless keyboards, mouses, and remote control. We have talked about some of the most notable ones that have paved the path to modern-day advancements. The reason you need to know about the past is because the past reflects the future. If you don't know what happened before, you will never have the idea what could happen next. Ordinary people in 1800 never even imagined that someday, people would be able to verbally communicate to someone who is across the world from them in real time. But here we are. However, there were some extraordinary characters who envisioned the future even in the 1800s. One such amazing individual was Nikola Tesla.

TWO

The Vision of Tesla

Would you believe it if someone told you that you can actually get from one part of the world to the other in seconds? Of course you wouldn't. It is impossible or at least it seems impossible when you think about it. Fifty years ago, people would never dare to dream how easily they would connect with people from all over the world and learn about whatever they are interested in. The advancement of technology in the last fifty years has made us believe that everything is possible, even wireless travel someday in the not-too-distant future. All technologies, wireless or wired, were the fruit of what some would call strange or irrelevant imagination. Some geniuses were so creative that they could see hundreds of years into the future. Such a genius in the field of wireless technology was Nikola Tesla, who was an American inventor.

Tesla: The Forgotten Genius

When we talk about the people who invented the greatest things, what name comes to your mind? Is it Thomas Edison or is it Henry Ford? I'm sure that there is someone whose name does not come naturally to you.

This is Nikola Tesla.

He died on January 7, 1943. His dead body was found in room no. 3327 of the New Yorker Hotel. Tesla had lived at this hotel for about a decade before his death at 86. He died without any kith or kin and with no money in his pocket. So how did one of the most prolific inventors the world had ever witnessed die in such poverty, with nothing to show for himself? Here is the story of Nikola Tesla.

Tesla was born in the town of Smiljan, present day Croatia, on July 10, 1856. When he was born, his mother called him the child of light. She did not know at the time that her words were nothing short of prophetic. When the little boy was five, his brother tragically died. He fell off a horse. This experience haunted him for many years. He started seeing visions along. His vision was also affected by light flashes that haunted him for the rest of his life. These visions later contributed to his many different inventions. He described how the designs were imprinted in his mind.

In his own words, Tesla said in the *Electrical Explainer*, "Invariably my device works as I conceived that it should and the experiment comes out exactly as I planned it. In 20 years there has not been a single exception" Tesla credited his prolific memory to his mother. His mother had a photographic memory and could remember most things she saw. Tesla used his powerful memory to understand engineering. His father wanted Tesla to become a priest. When Tesla was a teen and he had a bout with cholera. As he was very sick, the father promised that engineering school would be in his future if he survived. Guess what? He did survive! The young Tesla learned engineering at the Technical College of Graz, Austria, where he was highly regarded by all of his professors for his immense talent and passion for learning. Although he had tremendous potential, he could not finish school. He developed a gambling addiction, which forced him to drop out. He stayed away from all of his friends and family as well. That was the beginning of a series of unfortunate lows that Tesla encountered in his life.

Tesla eventually moved to Budapest and worked as an electrician to make money. One day, something strange and wonderful came to him. He came up with a way of developing a newer and better way to generate electricity. This could be done with alternating current. This invention would go on to greatly improve electricity generation. In 1882, he was employed by Thomas Edison's electric company. Although he started as a simple employee, his talent allowed him to quickly progress and get promoted. In 1884, he left Europe to work at Edison Machine Works in New York City. Edison became fond of Tesla very quickly, although it didn't last that long. As the relationship evolved, it soured, and the two men strongly hated each other. The two inventors disagreed on how electricity was supposed to work, especially how to transfer and store it. While Edison favored the use of direct current, the patent of which belonged to him, Tesla favored the use of alternating current. Alternating current is very important if you aim to have a steady electrical supply and can also provide much more power when distances are great between the source and the target. This is why to this date, all large appliances we use in our homes run on alternating current or AC, while the smaller devices like torches and batteries run on direct current. Edison's reasons for not taking up the use of AC because doing so could hurt his own patents. With subsequent disagreements between the two, Tesla decided to quit and go on his own, forming a company in the field in 1885.

The company did not quite work for him. Tesla lost his company and all his patents, as they were assigned to the company. This was another harsh blow to his efforts. After losing his patents, he was just surviving from month to month. But that did not last long either. Two years later, his luck changed when he invented a new motor. Running on alternating current, this was an induction motor. This was the best and most efficient motor at the time; it was used to convert power into mechanical energy. His great invention caught the eye of the investors at the American Institute of Electrical Engineers. One of them was

George Westinghouse, an investor who realized that this invention could be the ultimate answer to take out Edison's company. Tesla allowed Westinghouse rights to the AC motors' patent for a significant amount of money, royalties, and additional stock. He also worked for Westinghouse, acting as a consultant.

Thus began the war between Edison and Tesla. Edison's DC company was coming up against Tesla and Westinghouse. Edison's company tried its best to show people that AC was useless and even dangerous and left no stones unturned to discredit Tesla and Westinghouse. Despite that, Tesla and Westinghouse made great progress in their efforts with AC. They bid below Edison and were offered the contract for lightening the World's Columbian Exposition in 1893, which was held in Chicago. People who saw AC in action during this fair had no doubt that AC was the future. Tesla and Westinghouse also beat Edison when they built the first AV power plant at Niagara Falls. The plant was an outstanding success. It made Tesla the pioneer of renewable energy. Today, a statue of Tesla stands at Niagara Falls to remember his contribution.

This endeavor did not have a happy ending either. The Westinghouse company was losing money and went into serious debt. Tesla, having compassion for his friend and partner, gave up his royalties to help. It is assumed that Tesla gave up almost $300 million in today's money for the sake of Westinghouse's company. In return, as his company recovered, Westinghouse paid Tesla handsomely for the rights to use his patent. This gave Westinghouse the right to use it for as long as he wanted. Tesla started a new chapter in his life using this money.

He was finally financially independent and intended to set up multiple labs around New York for his projects. During this stage of his life, he made numerous inventions, including a version of light. This was a bladeless turbine that worked as a neon light. He also designed a remote control during this time. One of his standout inventions is the Tesla coil. Using this coil, you can achieve large amounts of electricity and send it to nearby devices wirelessly. This invention led to his ultimate vision of a wireless future, where we could transfer energy from one place to another without any wires, just like the Tesla coil. In 1895, a fire started in his lab and he lost years' worth of inventions and research all at once. The fire was the final blow to Tesla's ambitions.

When Marconi was experimenting with radio waves and applied for the patent, he was unsuccessful because Tesla was working on it at the same time. Later, Marconi sent messages between Europe and the US in 1901. The US Patent Office witnessed this and gave Marconi a patent for his invention. Tesla was furious, as Marconi used 17 of his patents to complete the work. He sued Marconi. This incident strongly impacted his mental health. This was a huge setback to his research into wireless communication. No investors trusted his instinct anymore. Investors were more inclined to fund Marconi's radio. So, Tesla was left in a disastrous state financially. He had to tragically give up on all his work and, from 1933, spent the last ten years of his life at the New Yorker Hotel. Westinghouse's corporation paid his living expenses. He did not need to pay rent, and yet still owed money to various people when he died. This was the end of the story of one of the greatest inventors of all time. Tesla was a genius who was ahead of his time and whose visions and dreams still live with us to this date. [1]

The Vision of Wireless Technology

As I mentioned earlier, Nikola Tesla was quite ahead of his time. He was not motivated by profits or money and capitalism did not appeal to him. His frail mental health could have contributed to his downfall to some extent. However, he had always wanted to enact change and he did it in many ways. But what was his last true vision? Why was he so obsessed with wireless energy transfer? And why couldn't he achieve it?

The number of inventions and patents that Tesla made and held is staggering. He invented motor induction, made the first hydroelectric power plant possible, and is also one of the fathers of radio technology. But among all his other inventions, the Tesla coil was the most special. With this machinery, it became possible to transfer energy from one device to another wirelessly. It made Tesla dream of a future where you could transfer energy from one spot to another without any wires. But as a source of power, the Tesla coil was not exactly practical. To light a large room, you'd also get electrocuted. Tesla wasn't able to fully comprehend the limitations of his invention; nobody did at the time. But the results made him imagine a greater future when a town or even a city can be fully electrified by using wireless energy transfer. His lab, which was known as the Wardenclyffe Lab, was the stage where his dreams became reality. A 57-meter tower was built based on the basics of the Tesla coil. He imagined that with this huge tower, he could transfer power to the whole city, just like a Tesla coil can electrify any device near it. He also envisioned using electricity from his lab for boats and planes. He also wanted his system to be able to broadcast information.

All of these were based on his belief that the ground and air are good conductors of electricity. He needed more towers to build a network that could supply electricity by zapping power into the ground and shooting it through the air. It also needed to be able to send messages and images just like radio waves. There is no doubt about the genius of Tesla's mind, but there were some factors that were missing from his knowledge. These facts were not yet established during his time. He believed air and ground to be great conductors of electricity, but we now know that they are not the best. In fact, the ground is a good insulator. You have to keep in mind that the experiments were all done before the discovery of atoms, so scientists really didn't know how conductors worked. You can say that Nikola Tesla was ahead of his time. Although he was building on amazing ideas, the base of his knowledge was faulty.

One Hundred Years into the Future

It has been more than one hundred years since Tesla's efforts at Wardenclyffe, yet when it comes to power, we are still very much using wires. The universal network of electricity that he imagined still hasn't yet arrived. Why is that? Humans have made many great revolutionary changes during this timeline. What was once considered impossible has become an everyday thing. Why haven't we yet achieved the vision of Tesla, though?

One of the reasons is that, although technology has developed a great deal, the laws of physics, conductors, and magnets have not changed. Remember that when you move away from the power source, its strength decreases. If you double down the distance, its intensity reduces by six folds. This is an issue of physics that you can't change. If you put it in context, the houses near the towers would be experiencing really efficient electricity, while the ones away from the power source would be inefficient. So, in this case, we are limited by physics. However, with the improvement of technology and materials, it is sometimes possible to defy laws of physics too. One example is the wireless cell phone charge we can do today; its mechanism is quite similar to the Tesla coil but on a comparatively smaller scale. It is practical and also efficient. If we did it, how come Tesla did not? It is because he didn't have the resources and the knowledge that we do.

Humans have scaled up the technology and now are looking for ways to expand it, such as charging an electric car wirelessly. This is possible through the use of a charging pad in a garage or in a rest stop. It is also possible to power small appliances with the use of magnetic fields. But there are potential downsides. In this case, a room needs to be outfitted with a metal skeleton. For traveling for greater distances, you need a medium other than electromagnetic fields. Scientists have done extensive research about transferring energy using ultrasound, microwaves, and even lasers. Although they all have merit, they can only be used on a small scale. All these methods are not quite appropriate or effective on a large scale because a huge amount of energy is wasted during its transfer from one carrier to another. To transfer kilowatts of energy, you will need megawatts of energy. This is not economically feasible. [2]

Tesla experienced some great achievements early on in his life. As he aged, he continued to do his thing but not as much as he did when he was younger. Eventually, he became isolated and forgotten by his peers and later passed away with no fanfare in the New Yorker Hotel. But his legacy still lives and so do his inventions. He is a hero to all physicists and engineers around the world. We often wonder what he would dream up if he was alive today.

THREE

Life Without Wireless Technologies

We have talked about how different wireless technologies have come into being and started to evolve. We are so used to these technologies that we cannot imagine a life without them. What would it really feel to live for a day, a week, a month, a year, or even your entire life without wireless technology? We would probably be more grateful than we are to the inventors, engineers, and scientists who have made our lives so much easier and comfortable using these technologies. They have also opened a wide variety of opportunities for earning money, learning, and socializing. How was life before the internet? How did people live? Were they happy? Is living without wireless technology impossible?

Use Your Imagination

Let us imagine for a while that you live in a time when there is no wireless technology. No wireless means no TV, no mobile phones, no radio, no internet, and no Wi-Fi. All these would be gone. Let's think about how this would affect you. Say you have an assignment to complete and you need to do it quickly. What is the first thing you do? Remember, you don't have your phone or a laptop anymore. You can't do research online. Your best bet is to go to a library where there are maybe hundreds of books on the topic you are looking for. You have to manually pick a book and look for the topic in question. While you narrowed your search to a few books, they don't cover all the information you need. And there's no search engine to look for more. If you don't understand a word or concept, you probably have to look for a dictionary to look it up. After a tiresome day spent doing research, you come back home and crave some entertainment. What is your best bet? Picking up a book, going out, or maybe playing outdoor sports. If you get injured while you play, you can't call 911 right away. No smartphone, remember? You or one of your friends will have to go to a telephone booth and call an ambulance.

I know the above doesn't sound that bad. In fact, picking up a book by your favorite author after a hard day sounds like a lot of fun. However, technology has made everything so much easier. We know its limits. If we didn't know how technology works, we'd be able to enjoy our lives just as well. Maybe we would be grateful for having a library, dictionaries, and telephone booths. However, if you were to be magically sent back to the 1980s, you'd find it hard. We can't forget the things we know, so it's hard for us to switch back to living in an analogue world. However, if you were born in the 17th century, for example, you'd probably have no idea that inventions like this were possible. You would try to stay happy with what you have, just as you do now. The difficulty is always relative.

How Hard Would Life Be?

Technology has transformed the world to the extent that nothing amazes us anymore. We keep wanting more and more. If your internet connection is bad for three minutes, you start complaining about your internet service provider. If your phone is slow, you get a new phone, probably the latest model. If your telecom is not giving you a good deal on your monthly plan, you switch. Without technology, you wouldn't be able to make these choices. They have also raised our expectations of what technology can and should be able to do.

When Nikola Tesla made a remote-controlled boat and controlled it wirelessly, people actually thought they were witnessing magic. The advancement of technology has raised the bar of both our expectations and

Wireless World

satisfaction. We get the newest model of a device, and less than a year later, we're looking for the next best thing. We have shifted the mindset from "This is impossible" to "This should be better." That is why we sometimes take technology, and notably wireless technology, for granted. The sole reason we would struggle to live without wireless technology is our mentality or mindset shift.

No TV and no radio technology mean that you would not be able to have access to any sort of entertainment and news other than what is found in newspapers and magazines. No TV means that you cannot watch movies, concerts, or your favorite team play. No internet means that you cannot look up any information and you can't learn online. You can't find the best tutorials for your hobbies and skills and you can't cook the recipe of your choice without a cookbook. It also means that you need to visit people in person to talk to them and you can't befriend people on social media. You'll need to send letters or fax to get in contact with family and friends who live far away. You'd be reading physical books, having to go to work every day, and using old rotary phones. No sharing of files, no listening to music on the radio, TV or YouTube, and no streaming of games would be possible. The list goes on and on. I'm sure that some of the sentences above surprised you because you take technology for granted. We are incredibly invested in wireless technology but we forgot how many things we're doing differently because of it. But now that you think of it, it seems incredible how different our lives would be without it. How would we be able to navigate life if all this information was taken away from us? It is almost unimaginable at this point.

The Influence on Human Behavior

The development of wireless technology is unbelievable and truly miraculous when you start to see how far we have come. But how much has human behavior evolved? Not so much. We are the same living beings that we were two hundred years ago; our cravings, our desires, and our wishes are quite similar. Wireless technology changed our extrinsic life but not our intrinsic behaviors. It has made our life easier than ever but it has not particularly made us happier.

Think about it for a second, when a person desired communication or attention before the advent of the internet, what would they do? They would visit other people at home for a nice chat or go to a pub or restaurant. These are some activities that you do today as well but in a different way because you're now using technology. You send a message to your friends or family or upload your party videos to social media. However, your behavior is still the same. If anyone wanted to gather knowledge, they went to the library. Now you probably just check out Amazon to find the book or read Wikipedia. If you wanted to send information to someone, you probably had to send it through snail mail. You now save a good amount of time by using email, but the end result is the same. Humans have always gained what they wanted, regardless of their access to technology. No matter how much technologies develop, our behavior remains the same it always was. People need company and connections. Humans influence technologies, not the other way. While technology constantly changes and develops, the most intrinsic needs and behaviors of humans are the same and will remain the same.

However, how we approach our lives can and does change because of technology. As things change, some parts of life you used to know disappear and are replaced with better technology. Think about pay phones if you can still remember them. As cell phones became the norm and smartphones replaced them and became affordable and ubiquitous, there was

Wireless World

no need for pay phones on every corner. This is how life changes as technology evolves.

The Life of an Amish

The world is filled with diversity. People of different religions, cultures, traditions, and colors live together in the society. The Amish are very different from other Americans. Never heard of them? The Amish are a group of people who follow a traditional Christian church originating from Swiss Germany. They reside in different places throughout the U.S. but the majority are found in Lancaster, Pennsylvania. So what is so special about the Amish people? And how are they related to wireless technology? Let's take a look.

The Amish people are unique and have a fascinating culture. They are known particularly for two things, simple living and avoidance of modern technologies. The Amish people migrated to Pennsylvania in the early 18th century. They are not allowed by their religion to use any kind of modern-day tech, including smart phones, TVs, and even cars. If they have a business that requires using lawn mowers or tractors, they don't drive these machines themselves. They also speak their own language, a mix of English and German. Higher education among the Amish people is discouraged; they are supposed to follow tradition. This is because they believe it might lead to social segregation and also the loss of their community. Their professions include farming, barn building, wood works, cattle rearing, and more. They live in rural settings and provide manual labor to get by. They try their best to avoid contact with outsiders due to cultural and religious reasons. It is amazing to see this kind of culture still alive in a world that is constantly trying to increase access and advancement of technology. We are all trying to urbanize our life as much as possible. Taking a close look at Amish people is an eye-opening experience about how other people live.

The reason we are talking about the Amish people is because they are proof that people could live quite well without technology. Unlike them, we have built a mindset that if we don't have the latest set of gadgets, we are missing out on life. The Amish people are a piece of inspiration who are happy with their own culture and lack of access to technology. They don't care about fancy gadgets and modern tech, yet they are living life to the fullest. They are possibly enjoying and living life debatably more than we do. Sometimes, technology puts us in a bubble, a bubble of not being good enough, not having enough, not being happy enough. The Amish people are the complete opposite of that. They manage to survive and lead a comfortable life outside of the technology bubble.

We have come to a time in the modern era where we can't imagine a life without technology; but everything depends on our mindsets. There are numerous people around the world who still do not have access to the latest technology. Some people simply do not want its blessings because of their beloved traditions, like the Amish people. Their lives are not worse than ours, and could even be better in some respects. We are so interested in technology and place so much value on it that our lives seem quite impossible without it. But in reality, humans were meant to survive no matter what, with or without technology.

FOUR

The Growth of Wireless Technology

Since its very invention, wireless technology was considered a revolutionary invention that would change the course of history. It certainly did that. It has an immense impact on our lives and lets us stay connected with the whole world. Its' growth has led to the improvement of so many aspects of our lives, including business, personal lives, marketing, and education, among others. The growth of these technologies is far from ending. It's not diminishing anytime soon either. Scientists, inventors, marketers, educators, entrepreneurs, and other professionals are trying to make the full use of the technologies we possess. Although wireless technology has its downsides, which we will be talking about in the next chapters, its contribution to modern lives is incredible and not comparable with anything else.

Wireless Technology and Business

Money makes the world go round. Every technology on earth has helped to move the snowball of marketing and business forward. They have led to increased revenues and decreased production costs.

Let's talk about wireless communication. Just think about it, how easy is it to get informed about the latest product in the market? Or to get informed that a product has gone on sale? Or to provide you with discount codes that you can use? All these are examples of business that humans have been using for quite a while. Discounts, sales, and advertisements are not particularly new. They have been prevailing for a long time. But the way that they are delivered to people has changed and has become easier. Now, companies can send emails and text messages directly to your phone or computer. They can broadcast ads on TV and also through every social media out there. All of this can be done remotely without any physical contact or wires. In earlier times, people came to know about different products through newspapers, posters, TV ads, and banners. So, if you were not paying attention, or were reluctant to know about the world by reading newspapers, you probably had no idea which item was on sale and which was not.

Not only have wireless technologies made it easier to send and receive product information, but it has also expanded the market of the products. You can now order most things from anywhere in the world and have them delivered on your doorsteps in no time. Businesses are no longer tied to a location or a region. They can sell products anywhere, anytime. One visit to Amazon's website can help you to get anything from anywhere and you don't even have to bother bringing it home. It will be just left in front of your door. In earlier times, you literally had to think about a location to open your shop and you had to figure out which place would be the most profitable. What should you sell? With the advancements in technology, your answers to these questions couldn't be clearer. You may still have questions, but the answers are way easier to figure out. You can literally find out which product works for which type of clients and in which regions, all thanks to the internet.

Not only has wireless communication helped the companies to make more money, but it has also helped individuals in numerous ways. People can get jobs sitting in their home without physically being present in an office. You can work remotely in any company from anywhere in the world. With the advancement of wireless technology, there has also been a significant boost in efficiency. This has led to increased customer satisfaction, faster deliveries, and better collaboration between the seller and the buyer. In case you get a faulty or wrong product due to a delivery error, you can easily communicate with the seller and mitigate the mishap. Wireless communication excels in giving quality control along with maintaining steady business growth.

Wireless Technology and Education

Just like every other sector, wireless technologies have had an immense impact on education. As it continues to grow, wireless technology is coming up with newer and better ways to provide education to all individuals. To put it in context, the number of online degrees that are available has grown exponentially, with 30% of U.S. students taking at least one class online.[1] While online education is specifically aimed at non-traditional students, like working adults, it has also become pretty common among high-achieving traditional students as well. Online learning opportunities are accepted and valued because they offer interactive communication even without physically attending classes. Now this trend is being adopted by universities, high schools, and even kindergarten.

Introducing wireless technology to education has benefits for both teachers and students. As technology makes more software and hardware available, it is becoming possible to provide a completely immersive educational experience. From a student's perspective, it is a lot easier to attend classes remotely and learn on their own time. Geographical location will not be a barrier to education anymore. You can attend the lectures and seminars of the greatest professors and quench your thirst for knowledge from your own living room. Wireless education also helps in sharing information and knowledge throughout the world. You can share your wisdom with whoever you want from wherever you want. Although it has its potential downsides, which we will talk about in the next chapter, most often this is an ideal situation. Wireless classrooms or online classrooms also help you to partake in education despite your busy schedule because it doesn't require the time to travel to on-site classes.

From the teacher's end, it is easier to manage students wirelessly through online classes than traditional classrooms. Wireless learning allows teachers to consider the students' preferred learning styles while presenting lectures. Topics can be better explained by interacting through emails and chat sequences individually, which also takes less time. Teachers can also share the necessary textbooks and information with the students in the form of presentations. Wireless technology also breaks the barrier between teacher and student. It helps them to interact better because many students feel more comfortable speaking to their teachers from behind a screen than in real life.

Wireless Technology and Social Life

One of the key fields of wireless technology is communication. Wireless communication has opened vast possibilities for humankind. It has connected people of different societies, colors, cultures, and religions. Although having a connected world has cost us in many ways, it has also blessed us with numerous conveniences that we would never have otherwise. Starting from keeping in touch with your family members to trying to reconnect with your childhood best friends, maintaining relationships has become a lot easier. You don't have to send a letter to your pen friend who lives 3000 miles away anymore and they are not going to only read the letter three months later. You can now send your friend messages within seconds and he can literally reply to you within minutes. If you think of it, it seems ridiculous how slow communication was back in the day.

It is now possible to stay in constant touch with your family who are living far away from you. With the advancement of technology, you can actually see with your eyes what they are doing and how they feel even if they are a million miles away. In earlier times, parents did not get to see their children who went abroad to study or immigrated to other countries other than in pictures. Their desire to see their children in person was rarely fulfilled. Wireless technologies closed the gap between people and helped this desire to become reality. Parents now don't have to wait years and years to see the face of their children or their grandchildren. It has become much easier for families to stay connected and know more about each other because of wireless technology.

You can also arrange virtual gatherings where friends from all over the world can join and have a nice chat. Your busy schedule does not always allow you the luxury to hang out with our friends or to have a drink with your buddies in person. You rarely used to be able to reminisce about the beautiful memories you made with your friends. But through technology things have become easier. You can share old pictures, have a chat, and even get to see the friends that you have made over the years. It is a blessing that many people don't realize wasn't possible until not so long ago. Loneliness can be the root of depression and many other mental illnesses, but through social media, email, and instant messaging we all get to spend time with people we love.

Wireless Technology and Individual Lifestyles

It is often overlooked how much our individual lives have changed with the advancement of wireless technologies. They have opened the opportunity for numerous people to earn a living remotely and live the life of their dreams. People are now able to showcase their talent and work to not only their locales but also to the whole world. YouTube, Fiver, and Upwork are some of the top websites that have helped hundreds of people to earn a living. You can become famous, earn a living, and be respected because of your talent and effort without the need for physical presence.

The so-called gig-economy, of which the above-mentioned websites are a part of, would not be possible without wireless technologies. While not everybody wishes to be part of it, it is an important source of income for many people.

Not all talents get recognition everywhere. What is trash to someone can be gold to another person. Many individuals have found this statement utterly true while showcasing their skills through the internet and getting recognition from people all over the world. The world can now hear you when you have the talent to speak, they can listen to you when you can play the guitar, and they can appreciate you when you reach a milestone in your travels. You get a great opportunity to portray your individual talent and skills in front of the whole world. This is something that would be impossible without the internet and wireless technology.

Wireless Technology and Emergencies

Wireless technology has not only become our go-to for its convenience, but also its effectiveness in case of emergencies. From the beginning, wireless technology was able to help in minimizing or even preventing different emergency situations. It has enabled us to send information within seconds. This is especially important in emergency situations, like natural disasters, broadcasting public health information, political conflict, and more. What do you do when you get in a car crash? You call 911 on your cell phone or somebody else does it for you. This is the easiest example of how much each wireless technology contributes to our lives.

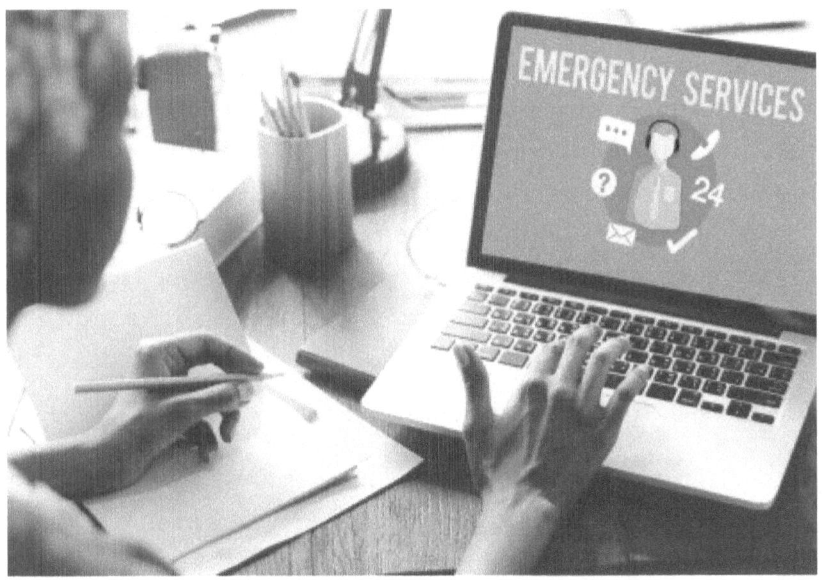

A cyclone is going to hit the coastal regions of a country. How do the people who live in the coastal regions know it? They found out on their radio, TV, social media, and local public announcements. All of these avenues are wireless. People can now know about the intensity of the storm in advance and are able to prepare accordingly. Thousands of people's lives are saved every year, thanks to the use of wireless technology. It cannot prevent natural calamities from happening but it can surely reduce the impact to human lives. What about the role of these technologies during a pandemic like COVID-19? Although many people get misled by misinformation (which we discuss in the next chapter), people can definitely stay informed by using these technologies. Wireless technology made it possible for information to get around the world. Within less than six months, people all over the world were alerted about the effects of COVID-19.

Here's another example. This is a real-life story about an Apple watch which literally saved a 67-year-old man's life in Norway. The man was supposedly wearing the Apple watch while accidentally fainting in his bathroom and hitting the floor hard. There was nobody around him to call the ambulance. But the fall detector on the watch, which detects when a person wearing it takes a hard fall, detected the fall. The watch both vibrates and makes a sound that can enable other people to help. It also shows an SOS button on the device's screen, so if a person is conscious, they can call for help by pressing the button. After a minute of no response from the wearer, the watch calls emergency services on its own. This is exactly what happened with Mr. Toralv. According to his daughter, he might have died if the Apple watch did not detect his fall and called for medical assistance. Mr. Toralv's life was saved. [2]

Technology has become so integrated in our lives that it is aware of our actions that it can save our lives. When we are in an emergency situation, we don't really realize how much we are dependent on these technologies. But their impact on our lives is more than we can acknowledge.

Wireless technologies have contributed to turning the whole world into a "global village." This is a place where everyone can stay in touch with anyone else in any part of the world. It is amazing how the concept of a global village is becoming more and more a reality as years go by. Wireless technologies have broken the barriers between cultures, traditions, and languages and helped to create a network that unites every single person in the world.

EMORY CLARK

FIVE

The Flip Side

Human civilization has flourished with the advancement of technology. It has played a part in developing each and every level of our lives, from the individual to the international level. It has improved manual labor and the ability to accomplish more things every day. But just like with every other thing in this world, there is always a flip side. Nothing in this world is perfect and this also applies to wireless technology. The flip side of the coin puts us in doubt about its risk/benefit ratio. Sometimes it also puts into question if we are actually heading towards an incredible future or utter destruction. Whatever it might be, you can protect yourself from all dangers with the help of mindfulness. The number one rule of staying safe is to know how something could potentially affect you and in what ways.

How Wireless Technology Affects the Body

A lot of research has been performed on wireless technologies to show their harmful effects. Although many specialists claim that their potential side effects are not dangerous, there are also many that do. So why is there a difference of opinion? Just like every other business, wireless technology has become a great source of revenue for different companies all over the world. Research that has been funded by these companies is likely to come up with positive results and tell you that you are absolutely safe to use their products. They don't care if you or your kids are exposed to teratogenic or carcinogenic elements; they just want to sell their products. This can cause intentional misinformation. However, people are exposed to radiation every day from their cell phones, TVs, Wi-Fi networks, and many other wireless devices. Radiation is likely to have negative effects on your body in the long run. Of 300 doctors who were surveyed about the effects of wireless devices, a staggering 99% agreed that they have a negative effect on human health. [1]

Many Wi-Fi studies have shown that it causes oxidative stress, testicular damage, psychiatric effects including electroencephalogram changes, cellular damage, endocrine defects, and calcium overload.[2] All of these effects can also be caused by other microwave frequency electromagnetic waves. Male infertility is one of the most alarming effects that can be caused due to the exposure to electromagnetic waves. Many men carry their smartphones in their pockets. This is very close to the groin region. The radiation from these smartphones can also change the shape, motility, and number of sperm. The major cells that get damaged due to radiation are the Leidig cells and the seminiferous tubules. The first produces the male reproductive hormone, testosterone, and the second is the region where sperm is produced, maintained, and stored. Exposure of human sperm to Wi-Fi internet increases the DNA fragmentation due to nonthermal effects. These types of wireless technologies should also be used with caution by pregnant women because of potential teratogenic risks.

The frequency range from 40MHz to 6GHz can penetrate tissue and has an impact on DNA. The radio frequency, known as electromagnetic radiation (RF-EMF), from wireless devices causes changes in the DNA shape and structure.[3] This may increase the occurrence of cancer and brain tumors. During the use of cell phones, radiation can penetrate up to 4 to 6 cm into the brain tissue. If cell phones are being used as little as one hour each day over 10 years or more, it greatly increases the risk for developing brain cancer. They are more likely to affect children rather than adults because their body and brain are still underdeveloped and in the process of development. Electromagnetic waves can also cause ear damage due to exposure to high frequency radiation. Longterm use of cell phones for four years or more for an average of 30 minutes each day is harmful to hearing. Young people who use their smartphones often because they are working in marketing and consulting services can experience ear pain and tinnitus. Microwave frequency EMFs are more dangerous for young children as their bone marrow cells remain in the hypercellular stage. The decreased repair and increased DNA damage by the exposure to EMF suggest that young children are getting more susceptible to cancer. The EMF action on stem cells can lead to the disruption of brain development, which is something that may cause autism in the future.[4] These are all very problematic issues and we can't rule out the causation of other problems as well. Exposing children to EMF radiation by using Wi-Fi networks in schools is, in view of all the above, not the best solution to applying technology in schools.

Other effects of EMF radiation even at low densities include headaches, dizziness, memory loss, and depression. These are some non-thermal effects that develop due to sensitivities of specific people to electromagnetic radiation. These phenomena are hard to measure and are still researched by scientists. In one study, researchers suggested that electromagnetic impulses can affect the nervous function of rats. They suggested that exposure to EMF radiation could contribute to dementia and Alzheimer's disease. More research is needed to determine if this is correct. [5]

Internationally, the guidance on radio frequency EMFs comes from the International Commission on Non-Ionizing Radiation Protection (ICNIRP). This is a non-governmental organization which is recognized by the World Health Organization (WHO). In 2020, WHO updated their guidelines for limiting larger radio frequency electromagnetic fields, which ranged from 100 GHz to 300 GHz [6]. It specifies that people should not work with higher impulses than 10mA m around their head, neck, and trunk areas. Any level above this is known to show radiation effects on the body and brain. Although most electrical appliances in our homes emit some amount of radiation, it is safe to say that most people are not exposed to excess EMF radiation.

Sitting in close proximity to some of these appliances can expose an individual to excess radiation, however. Remember that frequency fields may vary from one place to another, even within your house. So, you might want to check radiation readings throughout your house. Don't keep appliances close to your bedroom or that of your children.

Can Wireless Technology Contribute to Crimes?

Much like everything else, technology was created for the greater good but it can be abused. All technologies developed to this day were built for their benefits, not their disadvantages. But over time, humans find ways to manipulate the technology to commit crimes. This is when the question arises, does technology actually help us move forward? Wireless technology has opened multiple paths for people to engage in unsavory activities. As humans start becoming more and more inclined to use these technologies, cybersecurity and cybercrime are increasing. It is a shame how much potential wireless technology gives to would-be criminals.

Let's start with confidential data breaches. There are many companies out there that use weak or inefficient encryption methods due to their compatibility with older devices. These methods are easy to compromise and may lead to frequent hacks and loss of confidential information. Also, using the same pre-shared key for all devices can leave devices exposed to attacks.

This means that someone can set up an access point that broadcasts the same wireless network name as a trusted source. This process is known as a virtual honeypot or fake access point. You may think that you are connecting to a trustworthy network, like Starbucks' free wireless network, but you are actually connecting to an unknown third-party wireless network. Furthermore, if insecure authentication methods are used, where passwords are sent via text or in an easily decryptable hash, your information is likely to be breached.

The breach of confidential data is one of the most dangerous issues for any individual or company. No company would like to share their information, marketing strategies, and upcoming line of products with competition. A simple breach of confidential data can lead to a massive economic loss. With the advancement of wireless technology it has become possible to breach data from remote areas. Protecting information and data from hackers has now become one of the key concerns of every company, regardless of size. Even the slightest mistake can cost a company millions of dollars.

Personal security has also become a real issue. Hackers and scammers have developed elaborate ways to access your Facebook password, the pin of your bank account, and all the files in your computer. As the threat of cybercrimes continues to rise, International Data Corporation predicts that worldwide spending on cybersecurity will rise to approximately $133.7 billion by 2022.[7] Although companies try to do all they can to maintain online security for their users, sometimes the measures they take are just not enough. Every now and then, people get hacked and get their identities stolen. Hackers develop malicious viruses, malware, trojans, spyware, and ransomware that are able to steal your passwords and blackmail you. A report by Risk Based Security revealed that 7.9 billion records were exposed by data breaches in the first nine months of 2019 alone. The figure had more than doubled from the previous year. End-user protection is a key aspect of cybersecurity because often, users themselves accidentally help scammers and hackers to access their information by falling victim to phishing or uploading malware to their desktop or mobile device.

How Wireless Technology Impacts our Behaviors

Wireless technology has helped us to think ahead and look into the future. We mentioned the concept of a global village in the previous chapter and how the whole world is converting into a small village where everyone can connect with everyone. Wireless technologies have helped to defy all boundaries and distances. But just like every other thing in this world, wireless technology has impacted our mental health in ways that we don't often admit.

Happiness lies in contentment. But what is the biggest obstacle to achieving contentment? It is envy or jealousy or the desire that you need more to become happier. The concept of a global village is amazing but it has its downsides. Social media has created a virtual world where people post their happiest moments and their biggest achievements. When you scroll through different social media accounts, like Instagram, Facebook, and Twitter, you see all those smiling faces and achievements. This can affect you emotionally and mentally. Deep inside your mind, you start to feel that everyone is happier than you and everyone has something better than you do. Wireless technology has enabled us to see what everyone is doing and achieving but unfortunately it has not yet enabled us to see what a person is feeling inside. When you see a happy smiling picture, you don't see the person's struggle. Everybody struggles because nobody's life is perfect, no matter how rich and how famous they are. Homeless people and millionaires both have money problems; they're just different problems. When you look at them from the social media point of view, you see hollowness. This impacts your mental health negatively. We can't help but compare our lives to those of others. It is not logical. The people of whom we are envious feel the same way about us.

Wireless technology can also influence our emotions. Tech giants make good money by enticing social media users to click ads. Marketers use different human psychology concepts to make an impression on the minds of consumers. This is usually done in a way that you are unaware of how you have been manipulated. Advertisements are not limited to just radio or TV; they appear throughout social media. Their objective is to sell you a product and most often, these companies are successful. With the advancement of technology, it is no secret that companies like Google and Facebook track your online presence and behavior. They know what you like and what you talk about with your friends. All this information that we provide to them through using their apps gives them an upper hand in determining our shopping habits. The ads we see can seem completely random but they aren't. Everyone has a different kind of news feed based on their searches and conversations online. These companies can literally manipulate our needs and emotional attachment to specific topics and show us ads accordingly. This is a complete breach of privacy and also manipulation. More people should learn about it and talk about it. Wireless technologies have made our lives easier but they have also made our lives, our thoughts, and our habits more exposed than ever.

Wireless internet has helped humans to express their opinions and criticize others. But sometimes, these features of the internet are a doubleedged sword. Because of misinformation, many people now believe falsehoods that spread on social media. While some of these theories might sound bizarre, some people do believe them. The internet has made it possible to spread these theories across the world. Anyone from anywhere can come up with some imaginary scenario and upload it for the whole world to see. Some of these opinions are likely to harm people when they spread. Also, the existence of these social networks has led to many cases of cyberbullying. It might not seem like much of a threat but it is just as bad as getting bullied at school or at work. The comment sections of public groups and posts in different social media can be filled with hate and vitriol. These bullies do not realize how much a simple comment can affect a person's self-respect and self-esteem. These comments sometimes start as fun. The person who writes these comments doesn't really get to see or feel what the person at the other end is feeling. Internet trolling can sometimes have extreme effects on a person's mental health, and can even lead to suicide. Many people who take this extreme form of solving their problems have trolling to blame.

Is Wireless Technology the Biggest Distraction?

We talked about how wireless technology has become the backbone of our lives. We simply cannot live without it. It has made our life easier and communication faster. But how productively do we really use all these technologies? Does everyone benefit from these technologies? Or has it now become our number one distraction? The questions remain relevant.

Technological distraction might seem insignificant but it can come with heavy prices. Having unlimited access to information and entertainment can be a big problem for many. Research shows that it can take up to 23 minutes to fully return your full attention to what you're working on after being distracted. Just think about how many times you have picked up your phone while reading this book. You probably did it at least a couple of times. This means that every time you started reading you were not reading with full concentration, which may lead to poor retention. Students are particularly affected by this because the notification from their cell phones can distract them from an intense study session. The average attention span for humans has also decreased over the past two decades, all thanks to different social media and their ability to provide instant gratification. On average, the attention span lasts only eight seconds today. This is an alarming issue for everyone, not only students, because it is becoming easier to distract a person as we develop as a society. When a person becomes easy to distract, they also become easier to manipulate and that's just what some tech companies want.

Although wireless technology was built to save our time, at present, it has become more of a time waster. On average, people spend four to six hours each day looking at a screen and scrolling through it. With the increased number of TV channels, people often channel-surf without actually watching something. During large family gatherings, people are often more engaged in looking at their phone screens, or taking selfies than they are in interacting with relatives. It is safe to say that wireless technology saved us time but also contributed to wasting some of that saved time. We are far more interested in what someone on the other side of the world is doing than taking care of our own lives. We are far more likely to develop the desire for the newest gadgets instead of being grateful for the ones we already have.

The concept of having a global network shows us time and again that our life could get better. Our desires jump from one product to the other, our hobbies change weekly, our passion becomes boring work. All these things happen because technology influences us and our behaviors. Our values, our morals, and our wishes are all influenced by what we see on television and online. This can distract us from what we truly want and truly need. Wireless technology allows us to see that there is always something better to wish for, do, and live better. Technology was meant to take us closer to our goals. It was produced to help us pursue our goals easier. Don't fall prey to its dark side.

Let's get one thing straight. This book is not against wireless technology. The technology is amazing and its positive influence in our lives is incomparable. But we need to know where we are heading to. We have to know the downsides and the consequences of this technology. Only when we understand the threats and negative potential of this technology can we use it to do good. We must be aware of how these different technologies can harm us and take necessary protective measures against them. Wireless technologies were never meant to harm us. They were meant to help. It's important to remember that the number one rule for everyone is "Rule technology, or it will rule you."

SIX

Nature and Wireless Technology

By now, we have discussed at length how humans invented, evolved, used, and misused wireless technology. But we must not forget that there are other creatures living within this beautiful world. They are just as important in maintaining the ecological balance as we are. The human brain has given people the opportunity to become the best creation, but animals, trees, and birds are equally important for a green future. We have to think about the welfare of these natural elements of the world too. Do other living beings use any kind of wireless processes? Do these living beings feel the effects of wireless technology like we do? Do electromagnetic waves affect them the way they affect us? What potential effects could it cause to these living organisms in the long run? Let's take a look at what we know.

Animals Using Their Own Wireless Technology

Biology works in wonderful ways. It is amazing how humans can use electromagnetic waves, like radio and microwaves, but cannot see them with their naked eyes. Evolution has shaped us to see the bits of spectrum that are the most useful to see. Yet humans have found ways to use different kinds of EMF for different uses. What about nature? Can it feel the presence of different wireless waves around it? Can it use any kind of wireless waves to help it survive? The answer is not straightforward.

Not all animals have this feature to sense, feel, or see wireless waves around them. Most animals cannot see radio waves because that would require an impractical anatomy. There are always exceptions. As for UV sensitivity, many animals can see what is called human UV. Rod and cone pigments are sensitive to UV up to 280 nm. The UV sensitivity is higher in the visible range but humans cannot see it because wavelengths under 400 nm are absorbed by the lens and yellow pigment in the retina. Most animals have a UV-transparent eye medium, so they can see the human UV. Examples of such an animal are frogs or mice. As for human IR, many fish have cones with a maximum sensitivity of 620 nm. There are also reports that red sensitive photoreceptors exist in some butterflies, but they are not confirmed.

Most species of sharks have an additional sense known as the electroreception, the ability to sense faint electrical fields. This electric sensitivity allows them to detect mutual motion in water, fish, and the earth magnetic field. This would allow them to organize orientation and sense water movements and magnetic storms. This sensitivity is likely to be present in most living creatures with a nervous system but it is hard to say that they actually sense radio waves. They do have some sensation caused by the electromagnetic spectrum. Rattlesnakes and a few other species of snakes have what is called a "pit" organ, a thermal locator that helps them detect warm-blooded animals in complete darkness by sensing their thermal radiation. Pollinators like bees and butterflies can see ultraviolet light, which helps them to navigate to a flower's sweet spot. Pigeons are able to discriminate between almost identical shades of color; these are wavelengths that differ by very little. Pigeons can sense as many as five different spectral bands, unlike humans who have trichromatic color perception.

There is also the bio sonar, a feature that is commonly found in bats but also present in toothed whales, dolphins, and some cave dwelling birds. They emit high frequency sounds and use the returning echoes to form images of the environment. While humans mostly communicate with each other through sounds and sight, animals have developed wireless means of communicating with the help of smell as well. Pheromones are chemical smells that communicate everything from stress, danger, and fertility. Ants are the front liners when it comes to pheromones. Using pheromones, ants communicate with each other, warning each other about dangers, problems, opportunities, and more. Bacteria, insects, and mammals may also live or communicate with pheromones but humans are yet to discover what they are. [1]

Animals might not have invented fancy machinery and smart devices but there is every evidence of them using different wireless systems to communicate and sense the world around them. Through evolution, they have developed systems and organs that help them to survive. Even nature has found its way to integrate different wireless processes in different species to balance the ecosystem. There is still a lot that needs to be discovered about how animals perceive and use different wireless electromagnetic waves. Different animal features have helped humans to understand technology, including wireless technology, better. Humans are constantly learning from nature and the natural world. This knowledge can be applied in technology.

Effects of Wireless Technology on Nature

S tudies to evaluate the risks of radio frequency in biological systems have been conducted for quite a while. As technology advances, humans are looking for faster and more efficient ways of transferring data and energy. But these processes also need to be safe for humans and the environment. If they are not safe, then they are not going to be worth pursuing in the long run. Electromagnetic pollution caused by advanced machinery can be a possible source of risks for mammalian populations. It is important to conduct extensive research to determine the adverse effects of EMF waves on humans, as well as wildlife. Here is what we know as of now.

Some high-quality experiments on the biological responses of insects to naturally occurring electromagnetic fields demonstrate how they detect it and orient within it, as well as its effects on their behavior, physiology, and reproduction. Effects on insects tend to be complex, variable, and not always adverse. Of the few scientific laboratory experiments on some species, such as Drosophila, there is some evidence of cell damage, delayed development, and poor reproduction capability. There have been studies that have shown that close proximity to electromagnetic radiation from mobile phones can affect honeybee colony behavior. Exposure to mobile phone antennas can elicit idiosyncratic effects on pollination. In the U.S., the increasing influence of EMR pollution associated with a sudden decrease in the number of bees was observed a few years ago. This type of loss is known as colony collapse disorder (CCD). Affected bees cannot return to their hives due to failure of navigation. CCD has now spread to different countries, including the United Kingdom, Spain, Germany, and Italy. [2] Bees are an incredibly important part of the earth's ecosystem, even though many people do not realize that. Around onethird of the food produced today is pollinated by bees, making them a crucial part of the agricultural system.[3] They not only help with the food we eat but also the maintenance of the livestock's food. The effect of radiation on these smaller organisms might seem insignificant but their impact could be devastating.

In vertebrates, there is not much evidence that radiation has caused much harm in their behaviors or physiology. However, few studies exist on different species which are abundant near radio frequency towers or in cities with elevated EMR levels. The report findings range from altered hormone levels, problematic nociception, retarded growth, and malformations during embryonic development.[4] Of these, a reducing effect of repeated exposure to a zero magnetic field on nociception seems like the most established finding. In recent research conducted on bat colonies, behavioral changes were found in bats. They have also found morphological changes in amphibian species like increases in the heart rate, changes in blood count, and allergies.[5] Research indicates that EMF may have severe effects on animals, but more research is needed. In sum, the findings of radiofrequency influence on physiology of vertebrates are contradictory and inconclusive.

By far the most advanced research concerns the effects of radio frequency on the magnetic orientation of the migratory birds. Currently there is strong evidence that a protein in their bodies, called "cryptochrome," is naturally sensitive to magnetic fields in the radiofrequency range. It is proven that the innate compass in migratory birds can be disrupted by the weak frequency occurring in the larger cities. But the exact frequencies are yet to be discovered. Furthermore, some studies also claim that the fields from power lines affect the magnetic sense of vertebrates too. Mammals like bats and mice have a magnetic sense that may be disrupted due to the radio frequency emitted from different wireless devices. But it still remains to be discovered if the disruption of the innate magnetic sense in animals has massive ecological consequences or not.[6]

We have failed to shed light on one of the most important members of our ecosystems, the plants. Plants are like the backbone of the environment. It is vital to know if the advancement of wireless technology and excessive radiation from our devices can influence change in flora too. Several experiments conducted in labs tend to show that plant metabolism is affected by electromagnetic radiation. It is an established theory but still remains incomplete. However, the difference in the level of exposure makes it difficult to conclude what is happening in terms of plant metabolic processes. The impact of these changes is generally growth retardation but it is still hard to find a link between the changes and the plant's level of exposure since plant growth depends on many environmental factors. Electric fields with a strength above 20kV/m cause clear damage to the tips of plants. Other effects include delayed leaf senescence and leaf shedding in plants close to EMF radiating towers. However, effects are apparently of local and minor significance.

7

As you can clearly see, we are still behind in studying the ecological effects of EMF. The research on adverse effects of EMF must continue until we have better answers.

How 5G Affects the Environment

The fifth-generation wireless technology has been implemented throughout the world from the beginning of 2020. The technology comes with the promise of giving us faster data transmission. Although it seems really promising, its impacts on the environment need to be investigated. In this era where climate change and environmental pollution are getting worse, overlooking the impacts of newer technologies is very risky for the coming generations.

The main issue associated with implementing the 5G network comes with the development of 5G infrastructures. The new network is likely to cause a large increase in energy usage, which has already been one of the key factors for climate change. We have already talked about how wireless technology harms different living animals, including birds, bees, and plants. The 5G network is likely to increase how these species are affected, which may result in cascading effects throughout the entire ecosystem. The new 5G technology uses hardware that works with much higher frequencies, smaller cells, and massive multiple inputs/multiple outputs. These newer methods have expanded the potential of electronic devices. The way in which these technologies work together is complex, so we are not going to talk about them yet. The main component of 5G that may affect the environment is the millimeter waves. They have never been used at such scale before. This is why it is difficult to know their impacts on the environment and living organisms. A study by the Centre for Environment and Vocational studies of Punjab University observed that exposure to radiation from a tower for just under half an hour causes the eggs of sparrows to be destroyed. We have already talked about how radiation causes colony collapse disorders in bees; this effect is just going to get worse with the advent of 5G. [8]

Although government institutions like the Center for Disease Control and Prevention (CDC) and the Environmental Protection Agency (EPA) assure us that the radiation from these networks is entirely safe, several studies have been published that say otherwise. A group of more than 240 scientists have published several peer-reviewed papers on the effects of EMF and appealed to the UN for urgent action to reduce EMF exposure from wireless sources. Others have pointed out the FCC's RFR exposure limits consider the frequency of carrier waves but not the signaling property of RFR. The U.S National Toxicology Program (NTP) conducted research and found "clear evidence" that RFR exposure leads to increased DNA damage in rats and mice. A letter written by Dr. Martin Pall, a biochemistry professor at Washington State University, discussed the biological effects resulting from 5G. He also stated that the FCC guidelines are inadequate and favor the telecommunication industry. Dr. Pall also believes that there might be long term effects due to implementation of 5G, like hearing loss, skin cancer, and thyroid issues. The FCC has a completely different take on this matter. Instead of laying out effective guidelines, it puts efforts towards developing legislations that will prevent local governments from restricting the implementation of 5G. An open, unbiased, and frank investigation is needed on the subject and necessary steps must be taken to minimize the effects of 5G. We cannot let technology get in the way of environment protection and human wellbeing. That should be the first and utmost priority.

What Should We Do?

The EMF pollution in the environment has raised serious concerns around the world. Companies are trying to recognize how it can be minimized so that it doesn't harm us or our surroundings. The Faraday cage and Faraday fabric are some of the equipment whose invention indicates that it is possible for humans to at least create better protection against electromagnetic radiation at this time. A faraday cage creates a barrier between the internal component and the external electric field. It works because an external electric field redistributes the electric charges within the enclosure's conducting material, thus cancelling the field's effect in the cage's interior. This prevents the leakage of radiation from the equipment to the external environment. You can observe these effects in different appliances, for example, microwave ovens, coaxial cables, protective gear for linemen and electricians, and MRI scan rooms. All of these technologies use the concept of a Faraday cage to minimize the leakage of excess radiation into the environment. [10] Also, scientists have made faraday fabric a reality. They have created a faraday fabric by mixing ordinary cotton with a compound called MXene. This material is more of a broad category than it is a single compound and it is useful in lots of ways; this is just its most recent application. It has the ability to block electromagnetic inferences much better than any other materials. MXenes are conductive carbon compounds that can be fabricated in all forms, solid, liquid, and even sprays. A solution of tiny MXene flakes adheres to the fabric easily and blocks 99% of RF radiation in experiments. You don't need to cover yourself in it, however. Instead, you can make pockets or bags from similar fabric, which will help to minimize radiation from your phone or laptop to your body and the external environment. [11]

Extensive environmental studies are needed, since any adverse effects to animals, plants, birds or even insects can cause a great impact to the whole ecosystem. Radiation can ultimately impact human lives and human health, which would be a catastrophe. A key problem in this kind of research is the elimination of secondary sources of radiation, which is a difficult task. All epidemiological investigations should be double checked through experimental investigations.

Research has been spotty in these matters. A coordinated research agenda is required to address the issues of increasing use of EMF and its impacts on flora and fauna. It must be ensured that our ecosystem does not have to suffer permanently due to the advancements in wireless technology. Scientists should research the effects of man-made radiation on animals and plants more in depth. The appropriate choice of species is also very important in this matter. For example, birds are possibly the most exposed to radiation from wireless technology of all animals and they should be included in research. It is also necessary to develop effective guidelines for EMF exposure by drawing information from different established studies. Guidelines should be made, keeping both humans and the different habitats in which they live in mind. Another thing to study is whether radiation can be used to benefit the ecosystem. Scientists should research and investigate every possibility because the world is slowly advancing towards the ultimate vision of Tesla, which was a completely wireless future. If this proves to be a threat to the ecosystem, it is not worth pursuing.

Flora and fauna make up the balance of the world. They are required to maintain the ecosystem and protect the animals and humans from adverse effects. The advancement of human civilization has already done a lot of damage to nature. We don't want the environment to suffer even more for our convenience. That is why we need to take equal care of these animals and plants. As technology keeps progressing, the demand for information increases. Better wireless technologies ultimately lead to more EMF pollution in the environment. The level of EMF from different sources has risen exponentially and will keep on rising because of improvements in wireless technology. The current EMF levels that are due to the use of mobile phones keep increasing. We need to investigate and learn how EMF radiation can harm the environment and find ways to minimize its effects. We can't live without wireless technology, but we also cannot survive at all without the ecosystem.

SEVEN

Welcome to the Future

Wireless technology is constantly evolving. With the rapid growth of wireless technology over the last decade, planning for the future of wireless begins now. The addition of fifth-generation wireless is the beginning of a new era in this field. 5G wireless is super-fast and its mechanism of delivery is also different from that of its predecessors. Decreased latency and duplex data transmission provides amazing new opportunities for the future that are yet to unfold. Wireless communication has already overtaken wired technologies and will continue to expand in the future.

How Will It Progress?

The first question that comes naturally at this point is, "How will wireless technology progress from where we are at right now?" Is there still room for development? To answer simply, we can definitely say yes. Humans, who are naturally curious, always try to find a better and more efficient way to fulfill their desires through technologies and gadgets. Large companies are in charge of the changes in today's wireless industry. One of the main advantages of wireless technology is that it does not require a massive infrastructure of wired networks. This also opens the wireless technology market to smaller companies. Large companies are making efforts to invent better technologies that will help access the internet in areas that were hard to reach with wired technology. Facebook, Google, and SpaceX have all launched their own projects in this field. Facebook is working on drones that can fly at high altitudes to beam internet connectivity, Google is also taking a similar approach but with a different technology, known as weather balloons. Finally, SpaceX is working on low orbiting small satellites that can provide internet connectivity. These approaches will take wireless networks to the next level because there are still areas where the light of wireless communication and internet has not yet shone. About 52% of the world still lacks internet coverage and these efforts will help to provide access to rural communities and remote communities, especially in developing countries.[1] For the people who already have access to the internet the future of wireless technology will be different. Developed countries are seeking faster and more efficient ways to send and receive data. Companies are looking for advancements in different areas such as:

•Speed: The need for faster speed will never again be overlooked by the industry. In the future, wireless networks will be even faster than they already are. Increased speed will enable greater opportunities to enjoy your busy lives and schedules. **• Dependability:** While wireless technologies have become increasingly dependable and shortages are rare, they are expected to become even more dependable as the time passes. **• Security:** As we have previously seen, security is a great concern when it comes to wireless technology and devices. The security issues of wireless technology are pushing companies to seriously look to improve their devices in the future. [2]

Now let us take a look at what the future has in store for us.

5G and 6G

5G is available in most of the world, but it has not yet been implemented everywhere. It should be widely available within the next five years. Compared to previous generations, 5G will offer more speed. It will also decrease latency and improve the ability to connect to more devices. Speeds above 1 GB per second are expected to become a regular occurrence, with 10 GB per second also possible. 5G runs on two different frequencies, one low and one high. The low frequency 5G uses existing Wi-Fi and cellular bands, while high frequency 5G runs at frequencies between 28 and 39 GHz, which aren't yet used extensively. In addition to offering faster speeds, 5G is also crucial for advanced technologies like IoT and virtual reality, which require huge bandwidth and speed.

Even though 5G is still not available in all countries of the world, the industry is looking for the next big thing, which is 6G. This is because we are always future-oriented and driving for bigger and better things. The Center for Converged Terahertz Communication and Sensing (ComSenTer), which is part of the Semiconductor Research Corporation, is studying the technologies that will enable the 6th generation of wireless networks. ComSenTer says that 6G will operate at 100 GHz to 1 Terahertz (THz) frequency and offer up to 100 GB per second speed. [3] What this means is that you will be able to download almost an entire computer file within minutes. According to research, 6G will be able to connect hundreds or even thousands of connections simultaneously. The system is likely to be much more efficient than existing technology and will be able to offer more capacity with less power requirements. 6G is still in its initial stages and a lot of work still needs to be done. But hopefully this 6G technology could be in commercial use within the next ten years. You can see just how fast the wireless tech industry is evolving.

What can we expect from 6G? Well this is quite a hypothetical question because no one yet knows what it would be like to download a file at almost 1TB per second. It will be similar to 5G, yet better. Scientists expect 6G to go beyond a wired network; every device will act as a separate antenna using a decentralized network. If everything connects using 5G, 6G will push these devices to their limits and make data transfer instantaneous. While we are already expecting technologies like autonomous cards, drones, and smart cities that can use 5G, these developments will reach a whole new level with the advent of 6G. Some elements of what 6G would be able to do almost read like science fiction, such as the integrations of our brains with computers. Talking about the possibilities of 6G, NTT DoCoMo says, "It may be possible for cyberspace to support human thought and action in real time through wearables and micro devices mounted on the human body." Many have also called this the "teleportation of the senses." While this is not yet possible, it may well become a reality when 6G becomes available in the near future.

Li-Fi

This is another very exciting wireless technology that we may be getting in the future. Li-Fi is very similar to Wi-Fi, but it uses visible light waves instead of radio waves to send data. Researchers at University of Edinburgh have developed the technology and it has been tested in an Estonian factory. Specifically, it is a form of visible light communication system that uses light emitting diodes for data transmission. We already have Wi-Fi, so why do we need Li-Fi?

Li-Fi would be very advantageous for many reasons. It not only offers increased speed but also improved security. In experiments, it reached speeds up to 224 Gbps. In the test done in Estonia, it was able to hit about 1 GB per second. Currently, the development is focused on using LEDs to send data. Every LED light bulb could work as a wireless router in a Li-Fi network. The LEDs can be turned on and off so fast that we would not even be able to see them. This flashing will enable the LEDs to send data. They would flash in a Morse code-like pattern, which is made up of ones and zeros. Radio wave communication requires radio circuits. Li-Fi is actually simpler than that. It uses direct modulation methods which are similar to those of low-cost IR communication devices. One of the key benefits of Li-Fi is that there is no need to build any new infrastructure for it. It could work with the existing tech and infrastructure. Also, it can eliminate some of the security concerns inherent to wireless sharing. RF communication has always been vulnerable to eavesdropping and hacking. But signals emitted by Li-Fi remain confined within the space.[4] Li-Fi will allow us to eliminate these concerns and effectively protect our privacy. We have talked about how the world should move towards greener technologies. LED light bulbs are available anywhere, from homes, offices, and vehicles, among others, so if Li-Fi can be implemented, high-speed internet will be available wherever these lights are available. Although Li-Fi is being used currently in several countries, it is not yet effectively used all over the world.

Wireless Charging

Wireless charging is already a thing and you can wirelessly charge your phone, earbuds, smart watch, and more. The technology has now found its greatest range of applications. but it is still in its early stages. Wireless charging technology is quickly evolving and improving. Current technology is not really wireless because you still have to plug the charger in the electrical socket. You can then put your device on the charging mat or plate without having to plug it in. The device still needs to be in contact with the charger though.

In the future, we will be able to charge our phones without any physical contact with another device. Large main power sources could be placed on rooms ceilings and help charge multiple devices at once. Multiple companies are working on innovating this type of charging. This will be the closest thing to Tesla's original vision of electricity. Energous, a company in this industry, has developed "wattup," a product that can charge devices from up to 15 feet away with a small central power source.[5] The automotive, healthcare, and manufacturing industries also want to embrace this technology, as it promises increased mobility and advances that could enable IoT devices to be charged from a distance.

The global wireless charging market is likely to grow to more than $30 billion by 2026.[6] With the increase of electric and hybrid vehicles, the industry is working to enable wireless charging for these vehicles. This could be the next big thing! In recent years, wireless charging has gained great momentum. Strategic installation of charging pads around cities and outside people's homes will prevent the hassle of plugging in your vehicle every single day. The transport secretary of the United Kingdom is taking great initiatives to revolutionize electric mobility.

The technology will allow multiple taxis to charge at once, which could be very convenient for the drivers. Future trends in wireless charging can be based on dynamic electric vehicle charging (DEVC). This technology will allow the vehicles to get charged while in motion. If this is successfully implemented, it would drastically increase the range for electric cars as they can be recharged during long road trips. Also, DEVC would eliminate the need for large energy batteries, resulting in more light-weighted automobiles. With the increasing impact of fuel driven cars on the environment, electric cars seem to be the future of automobiles. Wireless charging could be a key element in taking electric vehicles to the next level.

Another sector where wireless charging could be potentially useful is the medical appliances sector. Wireless charging can revolutionize the healthcare sector. The industry is already experimenting with different technologies, including smart wearables and other medical devices. This could change the sterilization and charging process of different medical devices, portables, and cart-based medical products.

Numerous cardiac devices, like pacemakers, use implanted batteries, while cardiac assist pumps use wires to power the device. In 2018, researchers from MIT developed in vivo networking (IVN), a system that allows implanted medical devices to operate by getting power from external radio waves. This system is an evolved form of wireless charging technology called mid-field coupling. Many scientists are experimenting with this technology to transfer power and data from radio waves to medical appliances. However, all applications until now have required an external receiver to transmit power to the device. The test device built by the MIT researchers was about the size of a rice grain.[8] It is believed that it could be made even more compact. This opens up a wide range of possibilities, like powerful implantable devices and microscopic embedded sensors. Physicians could implant hardware into patients that can help to monitor blood sugar and other biochemical parameters and wirelessly get information about a patient's health.

Wireless charging and similar technologies promise to expand the capability of wireless devices. The first generation of wireless charging still needs physical proximity but soon the distance at which it works could increase to more than 10 to 100 centimeters. Innovators will be able to create more and more applications for wireless chargers. Think about wireless charging in restaurants, offices, and kitchen counters. Think about coffee makers using wireless electricity to make your cup of coffee each morning. As technology continues to develop, in the not-too-distant future we will be able to wirelessly transmit power through the air just like Tesla originally envisioned it.

Wireless Electricity

Tesla's revolutionary idea to develop a technology that will enable us to get wireless power without the use of any cables still lives on. It has been about 100 years since he first thought about it, but now several companies are becoming interested in making Tesla's vision a reality. Companies like WiTricity and Energous have already built smaller devices that allow wireless charging. How close are we to being able to use wireless electricity?

Emrod, a government backed New Zealand startup, is leading the race with the world's first long-range, high power wireless power transmission. This is expected to replace copper line technology. The technology utilizes EM waves to transmit power over vast distances. Energy is converted to electromagnetic radiation and transmitted and received by antennas, just like other radio waves. The energy is then distributed locally by conventional means. The transmitting antenna converts electrical energy to microwaves and focuses it into beams. The microwave beam is sent through a series of relays until it hits another antenna, where it is ultimately converted to electric energy again. This technology is quite similar to that of radio technology, However, the amount of power generated in this fashion is huge.

However, the amount of raw energy sent via wireless power remains the biggest concern of this technology. To minimize the loss of energy during transmission, efficient ways must be developed. Emrod has found a way to approach this challenge. It uses concepts related to radar and optics. The company will use metamaterials to focus the transmitted radiation more tightly than earlier microwave-based attempts. This is thought to reduce power loss but it is still not as efficient as wired transmission. Emrod's CEO says it could be economically viable in some cases. The company is confident about improving its efficiency even more in the future. [9]

The technology is reliable because it will not be affected by weather or atmospheric conditions. This means that unexpected power shutdowns will be a thing of the past. There are still concerns about how safe this is for human health, however. Emrod is confident that its electromagnetic beam operates at frequencies which are not harmful to human lives.

You may be asking yourself why we need wireless electricity when wired electricity works just fine. Innovation is never a bad thing and improving existing grids is never a mistake. Besides, the cost of conventional methods of electric transmission is extremely high. Setting up infrastructure and maintaining it is also expensive. Also, wired electricity does not provide for the use of powerful electric lines in remote areas. Wireless electricity, first and foremost, will be a cheap and renewable energy source for the entire world.

In coming years, we hope that the efforts to further develop wireless technologies will be successful. This will shape the world into a better place. If only Tesla was still around to see his vision become a reality.

EIGHT

Wrapping Up

Wireless technology has kept evolving from the very discovery of electromagnetic waves to where we are now. It has been an amazing journey with lots of ups and downs, success and struggles, and inventions and drawbacks. Over much of history, wireless technology has been losing ground to wired technology because of the lower expense of the latter. But in the last ten years, everything has changed drastically. The advantage of mobility, allowing anyone to receive and transmit data on the go has changed the face of wireless technology. Today, wireless technology includes a multitude of devices, including garage door openers, baby monitors, smart phones, smart watches, robot vacuum cleaners, and many more.

The history of wireless technology is long and interesting. We have gone through all of it in the first chapter of the book. It really teaches us that even a tiny improvement over the years can lead to revolutionary changes in the world. Heinrich Rudolf Hertz, Guglielmo Marconi, Nikola Tesla, Philo Taylor Farnsworth, and many others have been the pioneers of the wireless technology we see today. Their struggles have not been in vain. While their resources were limited, they never lacked confidence. Their innovative instinct to create and discover led them to become legends in the field of technology and inventions. The world is forever grateful to these amazing geniuses who gave their all and showed us the path of this amazing wireless technology.

In today's world, it is unthinkable to live without wireless technology. Everything we do involves wireless technology in some way or the other; we chat online, text each other, shop online, and order our transportation means online. It is really hard to imagine how a week or a day could be without access to the internet. Technology was mostly made for saving our time and fulfilling our needs. Modern technology, however, has managed to evolve past that. It has a direct influence on how we behave and react. It has changed the way we think and how we live. Humans have not evolved into a new being in the last 100 years, but the way they communicate has changed dramatically. This was thanks to the wonders of technology, and especially wireless technology.

Wireless technology has enabled us to dissolve boundaries and develop a global village, which was once just a theory. Individuals, organizations, and nations can come together, connect with each other, and share their views. The interdependence between different organizations and nations has led to development and progress even in sectors that are outside the technology industry. In matters that affect all people equally, like national conflicts or environmental disasters, everyone from around the world is now able to raise their voice and say their piece due to the wide opportunities created by wireless technology.

One of the greatest things about technology is how it has become accessible and affordable for everyone. It is no longer the luxury that it once was; it is now a necessity. Scientists have found cheap and affordable ways to communicate, work, and live. With the concept of the internet of things (IoT), devices both perform their functions wirelessly and are connected to each other in a seamless fashion. The IoT has enabled us to integrate wireless technologies in places that we didn't expect, like our kitchen appliances, watches, refrigerators, doors, and curtains. For example, your phone can now tell the thermostat to adjust the temperature of the room before you get home and smart lights can turn on automatically when you enter a room. We were already living in highly evolved technology, but now it permeates every aspect of our lives. The increasing demand for speed and information is the reason we are developing so fast

Wireless World

in wireless technology. All the things that were unattainable a hundred years ago are now real. Wireless communication, wireless connectivity, and even wireless power are or will be possible. Soon, all that is around us, not just our phones, will be smart.

However, nothing is perfect. This also applies to wireless technology. It has made people lazy. We no longer need to work hard and build our own homes. We don't even have to get off your couch to order pizza. You can just pick up your phone and order in the pizza app. People still work, but they don't use their muscles and bodies to do it. This sedentary lifestyle is a reason for many chronic diseases.

Technology has also exposed us to the growing threat of data loss and identity theft. Hackers are more dangerous to our data than ever before. They misuse our data and sell it to the highest bidder. This information can be used to manipulate us, harm us, and otherwise threaten us. Although this is not something that is very common, the risk is always there. It is possible to minimize this threat by making our data more encrypted and blocking tracking of our information through ad blockers and other software.

With the increased development of technology, the radiation that wireless devices emit is also a risk. At the moment, scientists think that the radiation is too mild to cause mutation or damage to the DNA. Longer exposure to wireless devices can result in cumulative effects, especially in case of children. The tech industry should really start focusing on these issues of security and health because these are of utmost priority for any individual. The risk of causing cancer when in constant contact with wireless devices also needs to be researched better. There's nothing more important than health.

There is no doubt that the future is wireless. We have talked about upcoming technologies that are going to change the face of the world. These technologies will show the path for a newer generation of technologies. The time is not far away when you can find internet connectivity everywhere you go. With the advent of Li-Fi, it will be possible to make even the roads and highways connect to the internet. How amazing will that be?

5G is an amazing feat of wireless technology. It is interesting to see how people's need for faster connection grows by each passing year. Maybe it is in the curious and ambitious nature of humans to always look forward and never settle for anything but the best. This is definitely the case in IT. In the near future we are likely to see 6G networks that will provide even more speed and efficiency to human life. One of the most mind-blowing inventions is wireless charging and wireless electricity. Just imagine charging all your devices without plugging any cable in. That would be amazing. There are no boundaries to power and energy. Will 6G be the peak of wireless technology? Unlikely. I don't think we've even scratched the surface yet.

Either way, it's a good moment, I think, that you familiarize yourself with all the elements included in this book. This is because we're standing at a crossroads. We are going to see bigger and better things in the future, but also bigger threats. However, once we understand the history and uses of wireless technology, we're better prepared both for the amazing new developments that will come and any dangers that we're unaware of yet. This is the way.

Wireless technology has completely changed our lives. It has come as a blessing to all of us. But we should always keep asking questions and look out for potential problems and harm. We should not get carried away in either direction. The responsibility lies in our hands to use these amazing blessings of technology the right way. The development of wireless technology has made what used to be science fiction a reality. And we have only just begun!

References

___

Chapter 1

1- Wikipedia. n.d. "Photophone." Wikipedia. Accessed July 3, 2021. https://en.wikipedia.org/wiki/Photophone.

2-Wikipedia. n.d. "Wireless." Wikipedia. Accessed July 3, 2021. https://en.wikipedia.org/wiki/Wireless.

3-Grandmetric. n.d. "How does wireless work?" Grandmetric.com. Accessed July 3, 2021. https://www.

grandmetric.com/2018/03/01/explained-how-does-wirelesswork/.

4-Wikipedia. n.d. "History of radio." Wikipedia. Accessed July 3,

2021. https://en.wikipedia.org/wiki/

History_of_radio#cite_note-sparkmuseum.com-1.

5- Wikipedia. n.d. "History of radio." Wikipedia. Accessed July

3, 2021. https://en.wikipedia.org/wiki/

History_of_radio#cite_note-sparkmuseum.com-1.

6-Wikipedia. n.d. "Philo Farnsworth." Wikipedia. Accessed July 4, 2021. https://en.wikipedia.org/wiki/Philo_Farnsworth.

7 -Wikipedia. n.d. "History of television." Wikipedia. Accessed July 4, 2021. https://en.wikipedia.org/ wiki/History_of_television.

8- Android authority. n.d. "A quick history of Bluetooth."android authority.com. Accessed July 4, 2021. https://www. androidauthority.com/history-bluetooth-explained-846345/.

9- Eyenetworks. n.d. "The History of Wi-Fi." eyenetworks.no.

Accessed July 4, 2021. https://eyenetworks.no/en/wi-fi-history/.

Chapter 2

1-Theverge. n.d. "The century-long quest for worldwide wireless power." The Verge.com. Accessed July 5, 2021. https://www. theverge.com/science/22166050/tesla-nikola-wireless-power-acqi-charger-electricity.

2-Tesla Science Center. n.d. "History of Wardenclyffe." Tesla Science Center.org. Accessed July 5, 2021. https:// teslasciencecenter.org/history/.

Chapter 3

————

Chapter 4

1-US NEWS. N.D. "ONLINE Education May Transform Higher Ed." Usnews.com. Accessed July 6th, 2021. https://www.usnews.com/ education/online-education/articles/2011/04/20/onlineeducation-may-transform-higher-ed.

2-Gizmochina. n.d. "Apple Watch 4 Fall detector saves owner's life after he fainted and fell hard in the bathroom." Gizmochina.com. Accessed July 6th, 2021. https://www.

gizmochina.com/2019/02/05/apple-watch-4-fall-detector-savesowners-life-after-he-fainted-and-fell-hard-in-the-bathroom/.

Chapter 5

1,2,4 Science Direct. n.d. "Wi-Fi is an important threat to human health." Sciencedirect.com. Accessed July 8th, 2021. https://www.sciencedirect.com/science/article/pii/S0013935118300355
3-RSA, Larik, Suhag AK, and Larik FA. n.d. "Effects of Wireless Devices on Human Body." *Journal of uoJ Computer Science & Systems Biology*. Accessed July 8th, 2021. https://www. hilarispublisher.com/open-access/effects-of-wireless-devices-onhuman-body-jcsb-1000229.pdf.

5-Sciencedirect. n.d. "Long-term electromagnetic pulse exposure induces Abeta deposition and cognitive dysfunction through oxidative stress and overexpression of APP and BACE1." https://www.sciencedirect.com/. Accessed 8th, July. https://www.sciencedirect.com/science/article/ abs/pii/S0006899316301330.

6-The International Commission on Non-Ionizing Radiation

Protection (ICNIRP). n.d. "ICNIRP GUIDELINES FOR LIMITING EXPOSURE TO ELECTROMAGNETIC FIELDS (100 KHZ TO 300 GHZ)." https://www.icnirp.org/. Accessed July 9th, 2021. https://www.icnirp.org/cms/upload/publications/ICNIRPrfgdl2020.pdf.

7,8-Risk based security. n.d. "Number of Records Exposed Up 112% in Q3." Riskbasedsecurity.com. Accessed July 9th, 2021. https://www.riskbasedsecurity.com/2019/11/12/number-ofrecords-exposed-up-112/.

Chapter 6

1- Researchgate. n.d. "Which animals can see the most extreme wavelengths of electromagnetic radiation." Researchgate.net. Accessed July 12, 2021. https://www.researchgate.net/post/ Which-animals-can-see-the-most-extreme-wavelengths-ofelectromagnetic-radiation-ie-whose-eyes-are-sensitive-furthestinto-the-UV-or-IR-spectrum

2,3,8-The Henry M.Jackson school of International studies. n.d.

"What Will 5G Mean for the Environment?"

https://jsis.washington.edu/. Accessed July 14th, 2021. https:// jsis.washington.edu/news/what-will-5g-mean-for-theenvironment/#_ftn39.

4,5,6 S. BATOOL, A. BIBI, F. FREZZA, and F. MANGINI. n.d.

"Benefits and hazards of electromagnetic waves,

telecommunication, physical and biomedical: a review." *European Review for Medical and Pharmacological Sciences.*

7-Knowledge and Learning mechanisms on Biodiversity and

Ecosystem sevices. n.d. "The impacts of artificial

Electromagnetic Radiation on wildlife (flora and fauna). Current knowledge overview." https://www.eklipse-mechanism.eu/. Accessed July 14th, 2021. https://www.eklipse-mechanism.eu/ documents/ 15803/0/EMR-

KnowledgeOverviewReport_FINAL_27042018.pdf/1326791cf39f-453c-8115-0d1c9d0ec942.

9-National Institute of Environmental Health Sciences (NIEHS).

n.d. "TOXICOLOGY AND CARCINOGENESIS STUDIES IN Hsd:SPRAGUE DAWLEY SD RATS EXPOSED TO

WHOLE-BODY RADIO FREQUENCY RADIATION AT A FREQUENCY (900 MHz) AND MODULATIONS."

https://www.niehs.nih.gov/.   https://www.niehs.nih.gov/ntptemp/tr595_508.pdf.

10- Leadertech inc. n.d. "What is a Faraday Cage?" https://leadertechinc.com. Accessed July 15th, 2021. https://leadertechinc.com/blog/applications-faraday-cage/. 11-Tech Crunch. n.d. "Here comes the Faraday fabric." techcrunch.com/. https://techcrunch.com/2020/12/11/herecomes-the-faraday-fabric/.

Chapter 7

1-UNITED PRESS INTERNATIONAL. n.d. "UN: Majority of

world's population lacks internet access." upi.com. Accessed July

16th, 2021. https://www.upi.com/Top_News/World-News/ 2017/09/18/UN-Majority-of-worlds-population-lacks-internetaccess/6571505782626/.

2,3,4,5-World Wide Supply. n.d. "Upcoming Wireless Network Technology." worldwidesupply.net. Accessed July 16th, 2021. https://worldwidesupply.net/blog/blogupcoming-wirelessnetwork-technology/.

6-Global market Insights. n.d. "Wireless Charging Market Size

By Technology (Inductive, RF, Resonant), By Application

(Automotive, Consumer Electronics, Industrial, Healthcare,

▪▪▪▪Aerospace & Defense), COVID-19 Impact Analysis, Regional Outlook, Growth Potential, Competitive Market Share &." www.gminsights.com. Accessed July 17, 2021. https://www. gminsights.com/industry-analysis/wireless-charging-market? utm_source=electronics-cooling&utm_medium=Editorial& utm_campaign=Sustainable%20and% 20Smart%20Technologies.

7,8-Electronics Cooling. n.d. "What Does the Future of Wireless Charging Technology Look Like?" electronics-cooling.com.

Accessed July 16th, 2021. https://www.electronics-cooling.com/ 2020/02/what-does-the-future-of-wireless-charging-technologylook-like/.

9-Okwatch, Leon. n.d. "How Far Are We From Wireless Electricity?" medium.com. Accessed July 16, 2021. https:// medium.com/predict/ how-far-are-we-from-wireless-electricity94dbd48529a4.